AF575658

AVIATION

P-38 Lightning, Vol. 2

Lockheed's P-38J to P-38M in World War II

DAVID DOYLE

Library of Congress Control Number: 2018956650

Technical Layout by Jack Chappell
Cover design by Justin Watkinson
Front cover photo by Rich Kolasa
Type set in Impact/Minion Pro/Univers LT Std

ISBN: 978-0-7643-5822-7
Printed in China

Published by Schiffer Publishing, Ltd.
4880 Lower Valley Road
Atglen, PA 19310
Phone: (610) 593-1777; Fax: (610) 593-2002
E-mail: Info@schifferbooks.com
Web: www.schifferbooks.com

For our complete selection of fine books on this and related subjects, please visit our website at www.schifferbooks.com. You may also write for a free catalog.

Acknowledgments

As with all of my projects, this book would not have been possible without the generous help of many friends. Instrumental to the completion of this book were Tom Kailbourn, Rich Kolasa, Dana Bell, Stan Piet, and Scott Taylor, as well the staff and volunteers at the National Museum of the United States Air Force. Most importantly, I am grateful for the help and support of my wife, Denise.

Contents

Introduction 004

CHAPTER 1 P-38J 006

CHAPTER 2 F-5B 034

CHAPTER 3 P-38L 038

CHAPTER 4 P-38M 076

CHAPTER 5 Combat 082

Introduction

While just over 2,000 P-38s had been built through the model P-38H, the largest group was the 1,082 P-38G models that were produced. While models P-38F through H were combat capable and indeed gave a good showing for themselves, with the advent of the P-38J the Army Air Force finally had the Lightning that they wanted. Powerful, long legged, and fast, 2,970 examples of the type were built. The P-38L model superseded the J and provided the Lightning with even more firepower through the addition of a rocket launchers, bringing with it the addition of a second manufacturer, Consolidated Vultee. In total, 3,810 examples were built by Lockheed, augmented by Vultee's plant in Nashville, Tennessee.

Visually, these aircraft are readily distinguishable from their predecessors by their distinctive air scoops below the propeller spinners.

The P-38 was one of the few aircraft to remain in continuous production throughout the US involvement in World War II. Because of this, it was one of the first fighters to go into combat in Europe with American pilots. Despite being initially badly outnumbered and facing problems with the early-model P-38s, ultimately over 130,000 missions were flown by P-38s in Europe, at a loss rate of 1.3 percent. In the Pacific theater, over 1,800 Japanese aircraft were shot down by P-38 pilots, with over 100 pilots earning ace status.

Once World War II ended, the P-38 was quickly phased out of service, with the military essentially standardizing on the Mustang until the ranks could be filled with new jet fighters. The USAF retired the last of its Lightnings in 1949. Today, only a couple dozen of the aircraft survive, and of these, only a handful are flying, with the rest displayed in museums; all keep alive the history of the twin-boom fighter.

Specifications

	P-38J	P-38L
Armament	4 x .50 cal. MG + 1 x 20 mm	4 x .50 cal. MG + 1 x 20 mm
Bomb load	4,000 lbs.	4,000 lbs.
Engines (2)	Allison V-1710-89/91	Allison V-1710-111/113
Maximum speed	414 mph @ 25,000 ft.	414 mph @ 25,000 ft.
Cruise speed	290 mph	290 mph
Service ceiling	44,000 ft.	44,000 ft.
Range	2,600 miles	2,600 miles
Wingspan	52 ft.	52 ft.
Length	37 ft. 10 in.	37 ft. 10 in.
Height	12 ft. 10 in.	12 ft. 10 in.
Weight	17,500 lbs. gross	17,500 lbs. gross
Number built/converted	2,970 + 200 F-5B	3,810 + 113 built by Consolidated

In June 1937, Lockheed received a contract to produce one XP-38 prototype. The overall lines of this aircraft, assigned US Army Air Corps serial number 38-326, are similar to the production P-38s, with some details differing, such as the lack of fixed oil-cooler intakes on the chins of the engine nacelles and the presence of air scoops on the tops of those nacelles. *National Museum of the United States Air Force*

The first P-38 model intended to be combat capable was the P-38D-LO, although in reality these aircraft failed to attain that status. They had a low-pressure oxygen system, a retractable landing light, self-sealing fuel tanks, and improvements to reduce buffeting and enhance recovery after a dive. This example with dummy gun barrels and the nickname "Snuffy" on the nose bears the red cross pertaining to the Red Force during the Carolina Maneuvers in late 1941. *Stan Piet collection*

The P-38E-LO was another step toward a genuinely combat-ready Lightning (although this model too failed to fully reach that goal). A total of 210 P-38E-LOs were completed. Some of the changes over the P-38D included the deletion of the large air scoops to the fronts of the turbosuperchargers, in favor of smaller scoops; a revamped nose landing gear with a smaller gear bay; more space for machine gun ammunition; and staggered .50-caliber machine guns in the nose. A 20 mm cannon also was included in the armaments package. *Air Force Historical Research Agency*

Beginning with the P-38F and continuing through the P-38H, shown here, the Lightning was truly a combat-ready aircraft. The P-38H differed from the P-38G principally in its more powerful Allison V-1710-89/91 engines, which were rated at 1,425 horsepower: a great improvement over the P-38G's Allison V-1710-51/55 (F10) engines, which were capable of only 1,325 horsepower on takeoff and 1,150 horsepower at 27,000 feet. The P-38H also had the M2C 20 mm cannon instead of the P-38G's M1, and its pylons had a higher capacity, 1,600 pounds on each one. *Air Force Historical Research Agency*

CHAPTER 1

P-38J

Due to its many external modifications, the P-38J had a markedly different appearance from the aircraft that preceded it.

Six blocks of the P-38J were produced; these ran from the P-38J-1-LO to the P-38J-25-LO. A total of 2,970 aircraft were built.

A redesign of the intercooling system had been needed for some time so the Allison V-1710-89 and V-1710-91 engines could perform at their maximum potential. Intercooling is vital for supercharged engines. It lowers intake air temperatures, which prevents predetonation (also known as knocking). This phenomenon can damage critical engine components.

The P-38J featured core-type intercoolers housed in a redesigned engine-nacelle chin. These had previously been mounted inside the leading edges of the outer wings. The intake for the oil coolers was located on either side of the intercooler intake. As a result of this redesign, space was created behind the leading edges of the outer wings. To take advantage of this, a fuel tank was now installed in the void, with a filler cap located on top of each wing. This modification began at about midproduction of the P-38J.

Other alterations included the installation of a hydraulic boost system for the ailerons, and the movable trim tabs were eliminated. The round control yoke in the cockpit was replaced with a new yoke with two pistol-type grips. Additionally, the cockpit heating and defrosting equipment was upgraded, and aircraft fuses were replaced with circuit breakers.

The previously installed curved windshield was replaced with a flat, bulletproof-glass front panel, starting with the P-38J-10-LO.

The compressibility problem that had been discovered with the YP-38 was eventually resolved with the P-38J-25-LO. Adding pilot-operated compressibility flaps just outboard of the engines solved this.

The P-38J finally created a truly combat-capable aircraft, and many prominent aces, such as Dick Bong and Tom McGuire, piloted it.

A P-38J-10-LO is undergoing a manufacturer's test flight over Southern California. Although the J-model Lightnings retained some of the features of its predecessors, the enlarged, flat-faced intakes on the fronts of the engine nacelles for the oil coolers and intercoolers were an entirely new development. Also, starting with the P-38J-10-LO, the bullet-resistant glass behind the windscreen was eliminated, and a new bulletproof glass panel became an integral part of the front of the windscreen.
Air Force Historical Research Agency

Lockheed P-38J-10-LO, serial number 42-67869, is observed from the left side. Faintly visible on the side of the nose is a stencil of the manufacturer's construction number, 2380. *National Museum of the United States Air Force*

The same P-38J-10-LO, serial number 42-67869 and construction number 2380, is viewed from the left front, showing the design of the new air intakes on the chins of the engine nacelles. On each chin, the center opening was for the intercooler air, and the outboard openings were for the two oil coolers. A polished, oval-shaped metal reflector was still on the inboard side of each engine nacelle, for monitoring the positions of the landing gear. *National Museum of the United States Air Force*

As seen in a frontal view of P-38J-10-LO, serial number 42-67869, the nose armament remained the same as on preceding models of the Lightning, with four .50-caliber machine guns and a 20 mm cannon in the bottom center position. *National Museum of the United States Air Force*

Lockheed P-38J-10-LO, serial number 42-67869, is seen from the left rear quarter, demonstrating the new, deeper shape of the chin of the engine nacelle. Although these larger chins resulted in more drag, the increased airflow to the coolers finally allowed the P-38J's engines to attain their full-rated performance at high altitudes. *National Museum of the United States Air Force*

A P-38J, serial number unknown, flies over an expanse of water. Lightning pilots appreciated the plane's twin engines, which sometimes could mean the difference between being lost at sea or returning safely to base when one of the engines was put out of commission during flights over the ocean.
National Museum of the United States Air Force

Lockheed P-38J-10-LO, serial number 42-68008, is undergoing a Lockheed factory evaluation flight over Southern California prior to delivery of the aircraft to the US Army Air Forces. The photo probably was taken in September or December 1943, during which months the P-38J-10-LOs were produced. *Air Force Historical Research Agency*

This is the first P-38J-15-LO, serial number 42-103979 and construction number 2813, prior to delivery to the Army. Under the wing is a 165-gallon drop tank. The pitot tube on the J-model Lightnings remained under the left wing. *Air Force Historical Research Agency*

On close inspection, the seeming puff of smoke to the front of the nose of this P-38J-15-LO, serial number 43-28859, may have been airbrushed on the original photo to simulate the firing of the guns in the nose. The P-38Js were powered by the Allison V-1710-89/91 engines. *Air Force Historical Research Agency*

Lockheed Service personnel are working on the right power plant of P-38J-15-LO serial number 42-104155. With the cowling panels removed, some parts of the engine, the outboard oil-cooler duct, and the plumbing are visible.

A factory-fresh P-38J-10-LO, serial number 42-68002, cruises above a suburban area. Drop tanks are visible below the wings, and antiglare panels have been painted on the upper inboard facets of the engine nacelles as well as on the forward part of the pilot's nacelle, or gondola, as the fuselage of the Lightning sometimes was referred to. *National Museum of the United States Air Force*

"YIPPEE," P-38J-20-LO 44-23296, was the 5,000th P-38 Lightning that Lockheed produced at the Burbank, California, plant, and to celebrate this milestone, the company gave the aircraft a flamboyant, bright-red paint job. In the cockpit is Milo Burcham, Lockheed's chief test pilot.

"YIPPEE" was painted on the bottom of the wings in large, white letters with dark-colored borders. This Lightning was rolled out of the factory in May 1944.

Standing on the left wing of "YIPPEE" at Lockheed's Burbank, California, facility on May 17, 1944, is chief test pilot Milo Burcham. "YIPPEE's" time as a brightly painted and prominently marked commemorative aircraft was brief, as the plane soon was sent to a combat squadron in the Pacific.

Contrasting with "YIPPEE's" red paint was the gray paint on her landing gear and the interiors of her landing-gear doors. After the stint in red paint and "YIPPEE" markings, the plane was sent to the 477th Fighter Group in the Philippines in June or July 1944.

P-38J-15-LO, serial number 44-23129, was converted to the prototype of the Pathfinder, a radar-equipped Lightning used for directing following planes in the formation during bombing runs. The plane had an AN/APS-15 BTO (bombing-through-overcast) radar set, with a large, bulbous radome located in the nose. A compartment with windows was provided for the radar operator to the rear of the radome.
San Diego Air and Space Museum

Lockheed P-38J-20-LO, serial number 44-23314, has been in airworthy condition for many years and currently is in the collection of the Planes of Fame Air Museum, in Chino, California, flying under the nickname "23 Skidoo." *Rich Kolasa*

P-38J-20-LO, serial number 44-23314, spent World War II in the United States. It served as a training aircraft at the Hancock Field School of Aeronautics, Santa Monica, California, in 1945. *Rich Kolasa*

In addition to the nickname "23 Skidoo," the plane also bears the name "Louise" on the left engine nacelle. The paintwork on the restored plane is Olive Drab over Neutral Gray. *Rich Kolasa*

Two Lockheed 165-gallon streamlined drop tanks are mounted on the pylons of "23 Skidoo." These tanks were made of stamped, low-carbon steel and were of semimonocoque design. Each tank included two half-shell stampings, which were seam-welded together. Lockheed ultimately was able to complete one of these tanks every 4.5 minutes. *Rich Kolasa*

"23 Skidoo" is viewed from the right-aft quarter while warming its engines. *Rich Kolasa*

The "23 Skidoo" name and nose art are not repeated on the right side of the pilot's nacelle. The yellow markings on the tail booms and the vertical tails are edged in white. *Rich Kolasa*

The retractable ladder is in the lowered position under the rear of the pilot's nacelle on "23 Skidoo." Visible on the tire of the right main landing gear is a diamond tread pattern. *Rich Kolasa*

Although "23 Skidoo" is generally a well-restored example of a P-38J, it varies in some respects from the aircraft as it appeared during World War II. For example, it is missing its cartridge-ejection chutes below the weapons bay in the nose. *Rich Kolasa*

"23 Skidoo" appears on a flight line with a cover secured over the cockpit windscreen and canopy. A Lockheed 165-gallon drop tank is visible below the left engine nacelle. *Photo by author*

The current nose art of "23 Skidoo" consists of a cartoon of a running Indian holding a tomahawk. Nine Japanese flags, replicating kill markings from the Pacific theater, are below the windscreen. *Rich Kolasa*

In a frontal view of "23 Skidoo," the fabric cover is still secured over the cockpit canopy. The landing-gear struts are painted a dull-aluminum color. *Photo by author*

With its engines running, "23 Skidoo" is observed from the rear on a hardstand at an airport. The high position of the pilot and the blister-type canopy gave the pilot an excellent all-around field of vision. *Rich Kolasa*

The National Air and Space Museum's Steven F. Udvar-Hazy Center in Chantilly, Virginia, preserves this P-38J-10-LO, serial number 42-67762 and construction number 422-2273. The US Army Air Forces accepted the plane on November 6, 1943, and a short time later Lockheed converted it to a two-seat trainer. By May 1944, this Lightning had been modified to P-38J-25-LO standards and was serving as a test aircraft at Wright Field, Ohio. In 1946, the plane was transferred to the Smithsonian's National Air Museum, predecessor of the National Air and Space Museum. *Photo by author*

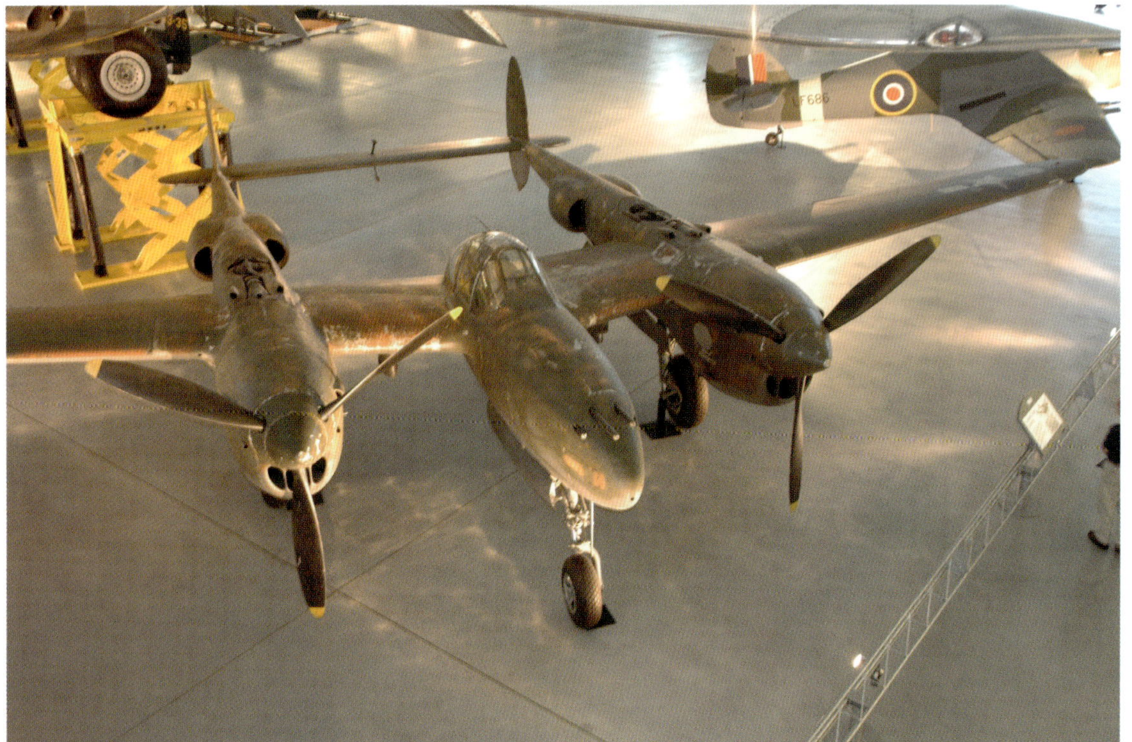

The National Air and Space Museum's P-38J-10-LO bears a well-worn original camouflage scheme of Olive Drab over Neutral Gray. *Photo by author*

The four .50-caliber machine guns are present in the nose of P-38J-10-LO, serial number 42-67762, but the 20 mm cannon was removed and its opening was faired over. On the tip of the nose is the round aperture for the gun camera. On the upper step of the leading edge of the left pylon is a square aperture for another gun camera. *Photo by author*

The nose landing gear is viewed from the left, showing details of the wheel, fork, oleo strut, torque link, wheel shimmy damper, and drag links. The shimmy damper prevented excessive oscillation of the wheel fork. A small stand is supporting the landing gear, for the protection of the 27-inch tire. To the front of the nose gear is a radio mast antenna. *Photo by author*

The left engine nacelle and propeller are shown. On the chin of the nacelle are the oil-cooler intakes; between them, but hidden by a propeller blade, is the intercooler air intake. The propeller blades have data stencils but not manufacturer's logo decals. *Photo by author*

The main landing gear of the National Air and Space Museum's P-38J-10-LO is mounted on a support. The oleo strut was painted gray, and there is considerable paint chipping, revealing the bare metal underneath. *Photo by author*

The oil intercooler shutter on the side of the left cowling is prominent in this photo. Above the shutter is an air intake fairing. On the bottom of the nacelle, the air outlet flap is open. *Photo by author*

The Lockheed-Fowler flap is lowered on the outer section of the left wing. The four flaps were interconnected, and they operated in unison with each other. They were fastened to carriages that rolled in tracks incorporated into the wings and were hydraulically operated. Also in view are the bullet-shaped carburetor-air intake and the outboard door of the left landing-gear bay. *Photo by author*

The left outboard Fowler flap of the National Air and Space Museum's P-38J-10-LO is viewed in its lowered position from the outboard side. *Photo by author*

A tinted, teardrop-shaped navigation-light lens is on the top and the bottom of each wing near the wingtip. This is the lower left one. The lenses of the left lights are red, and those of the right lights are green. *Photo by author*

The right wing and engine nacelle and the right side of the pilot's nacelle or fuselage are in view. Note the considerable chipping of paint on the propeller spinner and the leading edge of the wing. *Photo by author*

The right pylon lacks the stepped design of the front of the left pylon. On the inboard side of the engine nacelle, to the upper left, is an oval-shaped panel of bare metal. Originally, this panel was highly polished, forming a mirror by which the pilot could observe during flight if the nose landing gear was properly retracted or lowered. *Photo by author*

CHAPTER 2

F-5B

The new P-38J airframe's extended range and enhanced performance prompted Lockheed to use it as a platform for a more advanced series of photoreconnaissance aircraft. These became the F-5B-1-LO, F-5C-1-LO, F-5E-2-LO, and F-5E-3-LO.

The F-5B-1-LO was produced on the assembly line, and the other three models were converted. P-38L fighters served as the basis for the very similar F-5E-4. A Sperry autopilot was incorporated into the F-5B-1-LO and was the first Lightning to do so.

F-5B production was assigned Army serial numbers 42-67312 through 42-67401 and 42-68192 through 42-68301. Lockheed's designation for the F-5B-1-LO was model 422-81-21. Model designation 422-81-20 was assigned to the F-5C-1-LO; 422-81-22, to the F-5E-2-LO; and 422-81-23, to the F-5E-3-LO. All these aircraft were converted from P-38J production units.

The camera nose of the F-5B-1-LO was similar in shape to that of the F-5A. The arrangement of its camera windows was also reminiscent. Three camera bays were incorporated in the nose, and four different camera combinations could be installed. A 6-inch K-17 chart camera, 12- or 24-inch K-17 vertical camera, and 24-inch K-18 vertical camera could be selected. Typically, a K-17 6-inch oblique chart camera occupied the extreme forward position. Two 6-inch K-17 chart cameras or a single 12- or 24-inch K-17 chart camera were located in the center bay. Either a 24-inch K-18 or a pair of 24-inch K-17 reconnaissance cameras occupied the rearmost bay. In all, 200 F-5B-1-LOs were produced.

One hundred of the P-38J-15-LO aircraft were converted to the F-5E-2.

The US Navy was an operator of the F-5B, with four being deployed to North Africa. The Navy designation for the F-5B was FO-1. These aircraft were operated exclusively from land bases rather than carriers. However, Lockheed did conceive a carrier-based folding-wing variant of the Lightning—known as the model 822. This project did not go forward because of the Navy's preference for air-cooled engines.

The F-5B-1-LO was the photoreconnaissance version of the P-38J and specifically was based on the P-38J-5-LO airframe. Two hundred examples were produced. They had the same camera suite as their predecessor, the F-5A, and also featured a Sperry automatic pilot. The example shown here, F-5B-1-LO, serial number 42-67332, bears the standard Synthetic Haze camouflage paint. *Air Force Historical Research Agency*

Lockheed F-5B-1-LO, serial number 42-67332, foreground, flies in formation with P-38J-5-LO, serial number 42-67183. Both Lightnings wear the red-bordered national insignia used for several months in the summer of 1943. Note the light-gray exhaust staining around the turbosuperchargers on the F-5B. *Air Force Historical Research Agency*

Lockheed F-5B-1, serial number 42-67332, was representative of the first production block of photo-reconnaissance Lightnings based on the P-38J. These planes typically wore Synthetic Haze camouflage paint and were equipped with three camera bays, in which a combination of these cameras was installed: the K-17 chart camera, K-17 vertical camera, and K-18 vertical camera.

The same two Lightnings shown in the preceding photo, F-5B-1-LO, serial number 42-67332, foreground, and P-38J-5-LO, serial number 42-67183. The radio mast antenna on the F-5B was relocated from the bottom of the nose to the top. *Air Force Historical Research Agency*

CHAPTER 3

P-38L

More P-38L models were produced than any other variant of the Lightning family. A total of 3,924 were delivered.

Interestingly, from January to June 1945, Consolidated Vultee manufactured 113 P-38L-5-VNs in Nashville, Tennessee. These aircraft joined the onslaught of Lockheed's Burbank plant, where 1,291 P-38L-1-LOs and 2,250 P-38L-5-LOs were built from June 1944 to August 1945, bringing the total number produced to 3,811.

The P-38L mounted new power plants: the Allison V-1710-111 and V-1710-113. Each was rated at 1,425 horsepower. A feature that had been retrofitted to some earlier Lightnings was a gun camera. This was now incorporated into production and mounted in an extension at the upper front of the left pylon. This modification was necessary because there was less vibration compared to the camera's previous location in the nose, in close proximity to the gun muzzles.

Additional radio and tail-warning radar antennas along with compressibility flaps were added during production.

The F-5E, F-5F, and F-5G photoreconnaissance aircraft all were converted from completed P-38L aircraft.

Another converted aircraft was known as the Pathfinder. A test Pathfinder was converted from a P-38J, while all the others were modified from P-38L production. The Pathfinders carried a crew of two, the second crewman being the operator for the AN/APS-15 bombing-through-overcast (BTO) radar located in the nose. Most Pathfinders operated in the European and Mediterranean theaters, while the P-38L was primarily used in the Pacific theater.

The next model of Lightning after the P-38J, and the last model of production Lightnings, was the P-38L. The L-model planes were produced in the largest numbers, with Lockheed completing and delivering 3,924 of them. In addition, Convair assembled 113 P-38L-5-VNs at Nashville, Tennessee. This example, P-38L-5-LO, serial number 44-27112, is parked next to another P-38L-5-LO, serial number 44-26771. *Air Force Historical Research Agency*

In a frontal view of an unpainted P-38L, a feature introduced with the P-38L is visible on the leading edge of the left outer wing: the landing light was moved to this position from the bottom of the left wing, starting with the P-38L model. *National Museum of the United States Air Force*

A P-38L flies above clouds and an ocean surface. The small, dark shape of the lens for the landing/taxi light is visible on the leading edge of the outer left wing section.
National Museum of the United States Air Force

In honor of the first P-38L-5-VN, serial number 43-50226, produced by Convair at Nashville, a special inscription was painted on the left side of the nose: "VOLUNTEER / NASHVILLE CONVAIR'S / FIRST." "Volunteer" was a tribute to Tennessee's nickname, "the Volunteer State."
National Museum of the United States Air Force

An L-3 reflector gunsight is above the main instrument panel in this view of the interior of a P-38L cockpit. The gun-charging handle and selector knob, formerly a prominent feature to the lower left of the instrument panel in previous models of Lightnings, now was omitted.
National Museum of the United States Air Force

A slightly fuller view of a P-38L cockpit is presented in this photo. Toward the bottom of the photo are the control wheel and the top of the control column. On the upper left spoke of the control wheel is a small box marked "EXTEND DIVE FLAPS," containing the dive-flap control; this referred to the compressibility flaps added under the wings during P-38L production. *National Museum of the United States Air Force*

The P-38L cockpit is seen from the right. A new feature was the rocket-control box along the sill of the side window aft of the throttle quadrant and above the bombs and drop-tank control box. *National Museum of the United States Air Force*

"Christmas tree" multiple rocket launcher racks were introduced to Lightnings starting with the P-38L-5-LO model. A Lightning could carry two of these racks, each of which held five 5-inch air-to-ground rockets with high-explosive warheads. These rockets were highly effective against certain types of ground targets, such as tanks, strongpoints, and railroad trains. This example of launcher was mounted on P-38L-5-LO, serial number 44-25413. *National Museum of the United States Air Force*

Lockheed's Plant B-1 complex in Burbank, California, where the company produced the P-38 Lightnings, is shown in an aerial photograph taken in 1945. Empire Avenue is on the left, and the Southern Pacific railroad tracks are to the right. As wartime security measures, bushes have been planted for camouflage on the roofs of buildings, and camouflage netting has been rigged elsewhere. At the lower center are the three final P-38Ls, followed by one of the new Lockheed YP-80 or P-80A jet fighters.
National Museum of the United States Air Force

P-38L-5-LO, serial number 44-27231 and manufacturer's number 422-8235, was converted early on to an F-5G-LO photoreconnaissance Lightning. Since World War II, this plane has passed through a variety of private hands, flying under such nicknames as "Ruff Stuff" and "Scat III," in which guise it is seen in these photos. This Lightning flies under civil registration number NX79123. *Rich Kolasa*

Lockheed P-38L-5-LO, serial number 44-27231, bears the "LO" code for the 434th Fighter Squadron, 479th Fighter Group; the letter W is the individual aircraft's identifier. *Rich Kolasa*

"SCAT III" was named in honor of ace Robin Olds, who flew a succession of fighter planes nicknamed "SCAT" for the Air Force from World War II to the Vietnam War. *Rich Kolasa*

The flaps of the radiator housings are wide open in this right-rear view of "SCAT III" in flight. *Rich Kolasa*

Lockheed P-38L-5-LO, serial number 44-27083, has been restored to airworthy status under the nickname "Tangerine." It is in the collections of the Tillamook Air Museum, in Medford, Oregon. *Rich Kolasa*

P-38L-5-LO, serial number 44-27083, had been converted to an F-5G-6-LO photoreconnaissance Lightning but was restored to the P-38L configuration. Cartoon art of the singer and movie star Carmen Miranda (or a look-alike) adorns the right side of the nose. *Rich Kolasa*

Lockheed P-38L-5-LO, serial number 44-53232, is on display at the National Museum of the United States Air Force, Wright-Patterson Air Force Base, Ohio. It is marked to represent a P-38J assigned to the 55th Fighter Squadron, 20th Fighter Group, in England during World War II, with "KI" squadron code. *National Museum of the United States Air Force*

Three of the .50-caliber machine gun barrels with perforated cooling sleeves, and, below them, the 20 mm gun muzzle, are protruding through the nose of P-38L-5-LO, serial number 44-53232. A yellow border is painted around the left machine gun opening. *Photo by author*

The nose landing gear and the gear-bay door are seen from the front. At the top is the underside of the nose of the P-38L-5-LO. *Photo by author*

The nose gear is viewed close-up from the left side. Note the extreme compression of the oleo strut, and the nose wheel shimmy damper on the front of the strut above the torque link. *Photo by author*

The nose landing-gear bay of P-38L-5-LO, serial number 44-53232, is viewed from the rear, with the nose gear in the background. Attached to the strut toward the lower left are the two lower drag links, the tops of which are attached to the bottom part of the upper drag link. To the lower right is the interior of the nose landing-gear door. *Photo by author*

Part of the outboard side of the right forward boom are shown, along with a portion of the bottom of the outer right wing and the right landing-gear door. To the left is the carburetor air scoop. Each of the two forward booms extended from the firewall aft to the fronts of the radiator housings. *Photo by author*

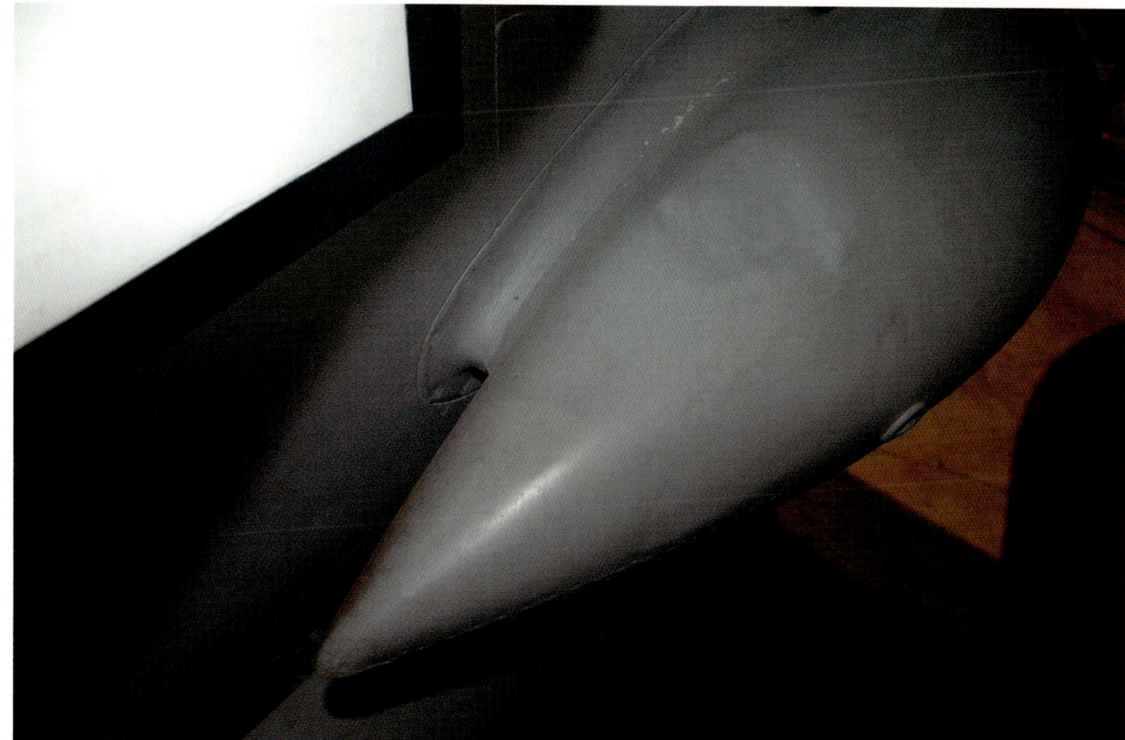

The right carburetor scoop is viewed from above, showing the shape of the flared, streamlined flange that serves to attach the scoop to the forward boom. *Photo by author*

The left carburetor air scoop is viewed from the front. *Photo by author*

The left pylon was of a different shape than the right pylon, with an extension on the upper front that housed a gun camera. A square aperture for the camera is nestled in an indentation on the leading edge of that extension. *Photo by author*

The left main landing gear of P-38L-5-LO, serial number 44-53232, is viewed from the front. The brake line is routed in a curve around the front of the antitorque link. *Photo by author*

The left main landing gear and the landing-gear bay are seen from the rear. The diagonal brace was called the side strut. The cast-metal crosspiece at the tops of the side strut and the oleo strut is the fulcrum. Each landing-gear assembly was operated by a hydraulic cylinder. *Photo by author*

The right turbosupercharger is in the foreground and observed at a low angle from the left rear. Each of the superchargers was driven by exhaust gases from the engine. Once the exhaust gases had passed through the supercharger, they were released through the waste gate, the duct with the open top to the far right. *Photo by author*

The right turbosupercharger is viewed from a very close perspective, to its right rear, showing the waste gate in the foreground. To the front of the turbosupercharger, nestled in an opening in the deck, is the exhaust pipe that feeds engine exhaust to the turbosupercharger. *Photo by author*

The cockpit canopy of P-38L-5-LO, serial number 44-53232, is observed from the right rear. Below the rear section of canopy are the radio transmitter and receiver units. Jutting from the top, hinged panel of the canopy is a rear-view mirror inside a Plexiglas housing. *Photo by author*

As seen from the front left of the cockpit canopy, the pilot's armor was removed from the cockpit, and the seat was fitted with leather upholstery. A clear view is provided of the X-braces on the roll-down side windows of the canopy. *Photo by author*

The cockpit of P-38L-5-LO, serial number 44-53232, is viewed from above. Stored on the left frame of the windscreen is a small spotlight. The control wheel is the late type, a butterfly shape with a grip on each side. The front of the windscreen was bullet resistant, fabricated from five layers of glass set in transparent plastic. The side panels were not bullet resistant and consisted of two layers of glass set in clear plastic. *National Museum of the United States Air Force*

Completed in June 1944, P-38L-5-LO, serial number 44-53087, has been restored and currently is in the collections of the Experimental Aircraft Association Museum, Oshkosh, Wisconsin. The plane was refinished to replicate the markings of fighter ace Capt. Richard Bong's famous P-38J, nicknamed "Marge." *Rich Kolasa*

The nose landing-gear oleo strut of P-38L-5-LO, serial number 44-53087, is seen from the left side, showing the wheel shimmy damper above the torque link "scissors." The lower drag links are the diagonal rods attached to the strut. *Photo by author*

The same elements of the nose gear shown in the preceding photo are viewed from the front. Details of the front of the nose gear door also are displayed. *Photo by author*

The interior of the nose gear bay door is shown. Attached to the rear of the door is its operating rod and cylinder. The door hinges also are visible. *Photo by author*

The P-38L-5-LO bears the "Marge" inscription and photograph as featured on Capt. Richard Bong's P-38J. The photo shows Bong's wife, Marjorie "Marge" Bong, in her college graduation photograph. *Rich Kolasa*

A view of the left engine nacelle, Curtiss Electric propeller, and spinner also includes the landing-light lens, which is curved to match the leading edge of the left wing. *Photo by author*

The left propeller, spinner, engine nacelle, and wing are observed from the front. Also in view in this photo and the preceding one is the cockpit canopy, including the windscreen with its flat, bulletproof front panel. *Photo by author*

A close-up photo of the front of the left engine nacelle provides details of the air scoops for the oil coolers (the two outboard scoops) and the intercooler (center scoop). The yellow stencil on the propeller blade provided its drawing number, serial number, and pitch data. *Photo by author*

The outboard side of the left engine nacelle and tail boom are depicted, including the carburetor air intake and, aft of it, the outboard radiator housing. The blister on the bottom of the wing toward the upper right was the fairing for a fuel booster pump, a feature introduced with the P-38L-5-LO. There was one such fairing under each outer wing; a flat, hinged access door was mounted on each fairing. *Photo by author*

The bullet-shaped object on the tail boom below the trailing edge of the wing is the carburetor air intake. Above that air intake, on the top of the tail boom, are elements of the turbosupercharger. *Photo by author*

The pitot tube and its mounting strut are located under the outer section of the left wing. The top of the mounting strut has a flared flange, which is welded to a square mounting plate that is fastened to the wing with flatheaded Phillips screws. *Photo by author*

A close-up photo of the outer section of the left wing includes, *left*, the raised aileron, and, below the wing to the right, the left compressibility flap in the lowered position. The compressibility flaps, one of which was under each wing, were introduced with the P-38J-25-LO to prevent the shock wave that was generated over the wings during high-speed dives, which disabled the operation of the elevators. *Photo by author*

The landing light, the pitot tube, and part of the left compressibility flap are present in this image. *Photo by author*

As seen in a left-front view of the left main landing gear, the angled tube to the front of the oleo strut is the drag strut, which was operated by a hydraulic cylinder inside the bay. The other angled tube is the side strut, which reinforced the oleo strut. *Photo by author*

The left main landing gear is observed from the rear. The top of the oleo strut is tapered, and it is attached to a crosspiece called the fulcrum, which in turn is attached to bearings on the sidewalls of the gear bay. *Photo by author*

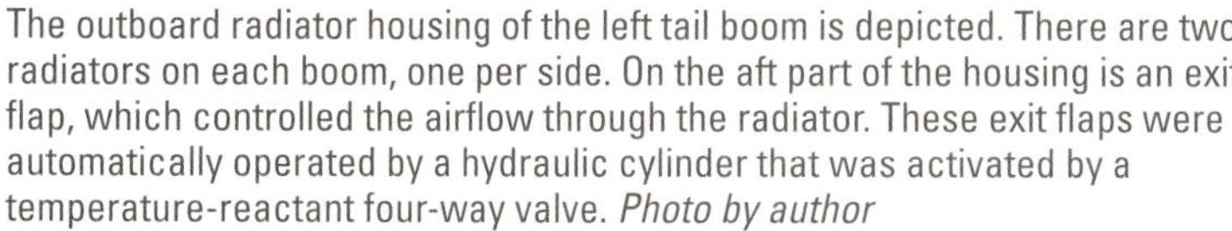
The outboard radiator housing of the left tail boom is depicted. There are two radiators on each boom, one per side. On the aft part of the housing is an exit flap, which controlled the airflow through the radiator. These exit flaps were automatically operated by a hydraulic cylinder that was activated by a temperature-reactant four-way valve. *Photo by author*

The red-painted left outboard extension of the horizontal stabilizer of P-38L-5-LO, serial number 44-53087, is viewed from the side. Note how the rudder (*right*) is notched around the stabilizer to allow the rudder to move freely back and forth. *Photo by author*

The left extension of the horizontal stabilizer is seen from a lower angle, along with the lower parts of the vertical fin and the rudder and the bottom hinge of the rudder. *Photo by author*

The entire left side of the left vertical fin and rudder are in view. Note the two prominent hinges for the rudder, the rudder trim tab, and the various access panels on the fin. *Photo by author*

The left rudder, vertical fin, and parts of the horizontal stabilizer and elevator are displayed. A steel shoe was attached to the bottom of the vertical fin to protect the fin from damage in the event of a nose-high landing. *Photo by author*

The upper left part of the horizontal stabilizer and elevator are shown, in addition to part of the left vertical fin and rudder, the left tail boom and inboard radiator housing, and the rear of the pilot's nacelle and cockpit canopy. *Photo by author*

The rear of the inboard left radiator is visible inside its housing on the side of the tail boom. Below the radiator are the rears of the left main landing-gear bay doors. The rears of the doors have stiffeners with a single lightening hole in each one. *Photo by author*

The pilot's nacelle, as the fuselage was referred to, is seen from the rear. The feature on the fore-and-aft centerline of the nacelle aft of the canopy is the boarding ladder's latch and release handle. The boarding ladder is on the centerline of the bottom of the nacelle. *Photo by author*

The top and the bottom mass balances on the center front of the horizontal elevator are viewed from the right rear. There are two recesses on the right side of each balance for attaching it with nuts and screws to the streamlined, forward-angled supports. *Photo by author*

A view from above the left wing incorporates the tail booms, radiator housings, left turbosupercharger, and empennage. *Photo by author*

The cockpit canopy and part of the left engine nacelle and tail boom of P-38L-5-LO, serial number 44-53087. Markings replicating those on Richard Bong's P-38J nicknamed "Marge" are shown, including twenty-four Japanese-flag kill markings. *Photo by author*

The spinners of Lockheed P-38L-5-LO, serial number 44-53087, are painted glossy red. Below the bulletproof front panel of the windscreen, and positioned at about the same angle as that part of the windscreen, was a ¼-inch armor plate. The 20 mm cannon has been dismounted from this plane. *Photo by author*

Lockheed P-38L-5-LO, serial number 44-53087, is viewed from the right front at its home at the Experimental Aircraft Association Museum, Oshkosh, Wisconsin. A yellow tow bar is attached to the nose landing gear. *Rich Kolasa*

In a close-up view of the outboard side of the right engine nacelle, the large air scoop on the upper part of the cowling is the cooling-air scoop for the exhaust shroud, while the small duct to the lower front of that scoop is the spark-plug blast-tube scoop. *Photo by author*

The inboard side of the right engine nacelle also has air scoops for the exhaust shroud and the spark plugs. *Photo by author*

Protruding from the upper front of the left pylon of the P-38L-5-LO is an enclosure for the gun camera. This feature was not repeated on the right pylon. On the front of the enclosure is a small, square window for the camera lens. *Photo by author*

The right main landing gear is viewed from its inboard side. The brake line loops around the front of the torque link before being coupled to the 12.7-inch, nine-disc-type brake unit. Tires specified for the main landing gear were 36 inch, smooth contour. *Photo by author*

At the rear of the nose gear bay door is the hydraulic operating cylinder. Also in view are two of the door hinges. *Photo by author*

The device on the bottom of the right outer wing is the compressibility flap. Above it is a sticker warning ground crewmen of the danger of injuries from getting an arm or a hand caught by the flap should it snap shut. *Rich Kolasa*

CHAPTER 4

P-38M

This final variant of the Lightning was not actually a production aircraft. All eighty aircraft were converted from P-38L models at Lockheed's Dallas Modification Center. The P-38M was designed as a night fighter that could outperform the attack bomber–based P-70 or the very large P-61.

P-83L-5-LO serial number 44-25237 served as the basis for the prototype P-38M. The AN/APS-6 Airborne Intercept (AI) radar was mounted in a pod under the nose. The conversion also included a compartment for the radar operator, complete with a one-piece Plexiglas canopy, installed to the rear of the primary cockpit.

During testing there were numerous unsuccessful attempts at shielding the pilot's night vision from the glow of the hot turbosupercharger exhaust. A contract was issued following successful testing.

The contract specified that Lockheed's Dallas Modification Center convert eighty P-38L aircraft to P-38M configuration. The first completed aircraft made its maiden flight on January 5, 1945. Production included the one prototype along with the eighty operational P-38Ms.

The production P-38M was a unique-looking Lightning, receiving an overall gloss black paint finish at the factory. In spite of the modifications to the airframe and the roughly 500-pound weight gain caused by the additional crewman, the P-38M retained all the capabilities of the P-38L. Crews for the night fighter were trained at Hammer Field, California.

The P-38M night fighter was not a production model of Lightning, but rather a modification of eighty P-38Ls, carried out at Lockheed's Dallas Modification Center. The idea was to field a night fighter that was smaller and more maneuverable than the P-61 and P-70. This photo of P-38M, serial number 44-27234, shows several distinctive features of that model: the AN/APS-6 Airborne Intercept radar pod under the nose, and the extra cockpit with bubble canopy for the radar operator to the upper rear of the pilot's cockpit. *National Museum of the United States Air Force*

P-38M, serial number 44-27234, is viewed close-up. The last four digits of the serial number are stenciled on the nose. A fairing fastened to the bottom of the nose contained the mounting points for the AN/APS-6 radar pod. Conical flash suppressors were on the muzzles of the guns to protect the pilot's eyes from the blinding glare when the weapons were fired in the dark. *National Museum of the United States Air Force*

A radar operator and a pilot are seated in their respective cockpits in a P-38M. Just inches from the radar operator's face is his radar scope, with a cover over it. The operator had very little headroom under the bubble canopy. *National Museum of the United States Air Force*

"Night Lightning" and the number "6865" are painted in script on the nose of P-38M, serial number 44-26865. The P-38Ms were painted in gloss black overall. Tail numbers and nose numbers were in red paint. *National Museum of the United States Air Force*

Converted from P-38L-5-LO, serial number 44-27234, this P-38M is flying above foothills in California. The metal skin panels around the turbosupercharger were left unpainted.
National Museum of the United States Air Force

Lockheed P-38M, serial number 44-27234, is fitted with Christmas-tree launchers for 5-inch rockets. The polished-metal mirror on the inside of the left engine nacelle is prominent.
National Museum of the United States Air Force

CHAPTER 5
Combat

Normally, P-38s had only one pylon under each wing, but the 82nd Fighter Group, with the 15th Air Force in Italy, modified some of its P-38Js to have an extra pylon under each wing. This P-38J from the 96th Fighter Squadron, 82nd Fighter Group, has just released two 500-pound bombs, and two more 500-pound bombs are visible on pylons underneath the wings. *National Archives*

Army Air Forces personnel survey a shot-up P-38 piloted by 2nd Lt. Peter Dempsey from the 38th Fighter Squadron, 55th Fighter Group. Damage is visible on the right wing, aileron, Fowler flap, main-gear door, and radiator housing. *Roger Freeman collection*

Sgt. Julius Cooper, the aircraft painter of the 38th Fighter Squadron, 55th Fighter Group, is spraying paint on a P-38J Lightning at a base in England on April 11, 1944. In the right background, a gray drop tank is lying atop a stack of shipping crates. *Roger Freeman collection*

Personnel are congregated around a Lockheed F-5B, number 122, assigned to the 7th Photographic Reconnaissance Group at RAF Mount Farm, England, on April 22, 1944. Several publicity photographs were taken of this plane on this occasion, and a photographer with a press camera is standing on the wing. *Roger Freeman collection*

The most famous plane of the top-scoring US ace of World War II, Richard Bong, of the 9th Fighter Squadron, 49th Fighter Group, was P-38J-15-LO, serial number 42-103993, nicknamed "Marge," but Bong scored some of his kills in other P-38s, including this P-38J-15-LO, serial number 42-104380. Bong flew this Lightning while based at Nadzab, New Guinea, around April 1944. *Stan Piet collection*

Ground crewmen are preparing for the next mission of P-38J-15-LO, serial number 42-104207, from the 55th Fighter Squadron, 20th Fighter Group. This Lightning bore squadron/aircraft code KI-N. On July 16, 1944, this plane was lost in combat, but the pilot, Lt. John Klink, survived, escaping through enemy lines to safety. *Roger Freeman collection*

Nicknamed "Wrangler," P-38J-15-LO, serial number 43-28393, wore this black triangle on its tail while serving with the 55th Fighter Squadron, 20th Fighter Group. *Roger Freeman collection*

An armorer with a belt of .50-caliber ammunition draped over his shoulder poses next to "Wrangler," P-38J-15-LO, serial number 43-28393, in or around May 1944. Later, while serving with the 474th Fighter Group, this Lightning and its pilot, 2nd Lt. Jack W. Haggard, were lost in a crash near Strasbourg, France, on June 8, 1945. *Roger Freeman collection*

Lockheed P-38J-10-LO, serial number 42-67916, was nicknamed "California Cutie" while assigned to Lt. Richard Loehnert, a pilot with the 55th Fighter Squadron, 20th Fighter Group. Dozens of mission markings are painted in yellow and include locomotive engines, top hats, umbrellas, brooms, and bombs. *Roger Freeman collection*

Three armorers wearing sun helmets are mounting a three-tube AC-MIC 4.5-inch rocket launcher combination on the lower quarter of the central nacelle of a P-38J-10-LO assigned to Maj. Willard J. Webb of the 459th Fighter Squadron. The scene was an airfield at Chittagong, India, (present-day Chattogram, Bangladesh) during 1944. The fronts of the three tubes on the right side of the central nacelle are visible.

"The Florida Gator" was the nickname of F-5F-1-LO, serial number 42-67119, assigned to the 22nd Photographic Reconnaissance Squadron, 7th Photographic Reconnaissance Group, based at RAF Mount Farm, England. The usual kite-shaped window on the left side of the nose had been omitted. This plane and its pilot, Lt. Edward R. Durst, were lost when the plane failed to recover after a steep dive near Saint-Malo, France, on July 24, 1944. *Air Force Historical Research Agency*

The original label of this photograph states that it depicts the first Lockheed P-38 to land on an American airfield in France, in June 1944. In the background are a windsock and ships of an invasion fleet. This plane was P-38J-10-LO, serial number 42-68071, assigned to the 392nd Fighter Squadron, 367th Fighter Group, 9th Tactical Air Command. *Air Force Historical Research Agency*

The Lockheed F-5E featured a raised, teardrop-shaped fairing into which a window for an oblique camera was set, on each side of the nose. This example, numbered 311, has invasion stripes, the black-and-white recognition markings applied to Allied aircraft for the invasion of Normandy in June 1944.

Members of the 467th Service Squadron are working on the left engine of a P-38J from the 364th Fighter Squadron, 357th Fighter Group, 8th Air Force, at RAF Honington (USAAF Station F-375) on July 23, 1944. *National Archives*

Armorer Cpl. Joe Diaz loads a belt of 20 mm ammunition into the magazine in the nose of a P-38 nicknamed "Li'l Venus" in or before late September 1944. Note the masking tape wrapped around the gun muzzles, and the access plate that has been removed around the 20 mm cannon barrel. *National Archives*

In a September 1944 photograph, armorer Sgt. Sidney Simpkins, right, cleans the bore of the right .50-caliber machine gun barrel in the nose of a P-38 while the plane's crew chief, TSgt. Alvin L. Archer, inspects the ammo magazines. *National Archives*

With the screwdriver in his right hand, an armorer adjusts .50-caliber rounds in the feed chute of the right .50-caliber machine gun in a Lockheed P-38. The exposed portions of the barrels of the .50-caliber machine gun barrels of Lockheed Lightnings typically were covered by blast tubes. *National Archives*

Two armorers from the 94th Fighter Squadron, 1st Fighter Group, are loading ammunition into a P-38 Lightning at an unidentified airfield in Italy. On the side of the nose just aft of the nose cap is an access door for the gun camera. *National Archives*

The landing gear is being retracted during the takeoff of one of the 13th Photographic Reconnaissance Squadron's planes, F-5E-2-LO, serial number 43-28616, at RAF Mount Farm, England. On November 22, 1944, this plane was lost; the pilot, Lt. Allan V. Elston, survived but was interned in Sweden, apparently for the duration of the war.

A flight line of Lockheed Lightnings with the 36th Fighter Squadron at Hill Airfield, near San José, Mindoro Island, Philippines, on December 20, 1944, includes at least two P-38L-5-LOs: the second plane in line is serial number 44-25426, and the fourth is 44-25355. Many more P-38s are parked in the background. *National Archives*

1st Lt. Pius Kunz of the 397th Fighter Squadron, 367th Fighter Group, stands for his photograph next to his P-38J-15-LO, nicknamed "Napoleon's Delight'nin'," at an airfield in Belgium on December 24, 1944. Interestingly, the nose art is not painted on the aircraft but is on a sheet of material—paper or cloth—that has been taped to the airplane. *National Archives*

Lt. Col. Edwin Chickering of the 357th Fighter Group poses alongside his P-38L, nicknamed "Gung Ho," at an unidentified airfield in Belgium on January 3, 1945. The nose art involved caricatures of "Bugs Bunny," an angry dragon, and a bomb with a lit fuse. *National Archives*

Lt. C. R. Livingston, of the 367th Fighter Group, poses next to "Cock-Tail," a P-38J-10-LO, serial number 42-68004 and code H5-K. The photo was taken at a USAAF airfield in Belgium on December 11, 1944. *National Archives*

A pilot, *center*, is conferring with two ground crewmen dressed in shearling leather jackets and pants in front of a P-38 Lightning at a base of the 55th Fighter Group. The plane is painted in Olive Drab over Neutral Gray, and there are two kill markings above the aircraft data stencil on the side of the pilot's nacelle. *National Archives*

Ground crewmen are servicing an F-5B photoreconnaissance Lightning, numbered 273 on the nose, at an unidentified base in the winter of 1944–45. A notation on the photograph claims that the plane bears French insignia, and indeed a roundel does appear to be present on the left wing. Note the open, hinged front camera window on the bottom of the nose. *National Archives*

The two men next to the left engine nacelle of a Lockheed Lightning at a snowy airbase are preparing to hook up a portable heater to warm up the left engine. This is a P-38L, on the basis of the presence of the landing light in the leading edge of the left wing. *National Archives*

Mechanics from the 9th Air Force are performing a major overhaul of a P-38J at a base in Belgium in the final months of World War II. They have removed the guns, the left engine, the structural frame of the engine nacelle (except for the engine mount), and much of the aluminum-alloy skin of the pilot's nacelle. *National Archives*

In another photo taken at an airbase in Belgium late in World War II, mechanics are doing maintenance or repairs on a P-38J-5-LO nicknamed "Mary Rose II," assigned to the 485th Fighter Squadron. A 6 x 6 mobile maintenance shop truck is backed up to the plane, availing the mechanics of compressed air, tools, and shop facilities. *National Archives*

9th Air Force mechanics are repairing both of the Allison engines of a well-weathered P-38 at a location in Belgium. An engine is awaiting disposition in the left of the photo. With the propellers and spinners removed, the round frontal plates of the nacelles are visible. *National Archives*

Mechanics brave cold weather in Belgium to make repairs to the engine of a Lockheed Lightning. Above the head of the second man from the right are the exhaust manifold and the black-colored left valve cover. *National Archives*

The F-5E-4-LO was a photorecon version of the Lightning based on the P-38L airframe. This example, numbered 225, served with the 7th Photographic Reconnaissance Group at RAF Mount Farm, England, and it exhibits the white rudders of the 22nd Photographic Reconnaissance Squadron. *Stan Piet collection*

This F-5B-1-LO, serial number 42-68205, assigned to the 7th Photographic Reconnaissance Group at Mount Farm, also sports the white rudders used on aircraft of the 22nd Photographic Reconnaissance Squadron. *Roger Freeman collection*

The "Droopsnoot" Lightnings were P-38Js and P-38Ls modified with clear noses for a bombardier in the forward compartment. He was equipped with a Norden bombsight and was able guide large formations of Lightnings precisely onto a target and signal them when to drop their bombs. Lockheed engineers developed the Droopsnoots at British bases. *National Museum of the United States Air Force*

A Droopsnoot Lightning is viewed from the front, showing the shape of the clear panel on the lower front of the bombardier's nose. The bombardier's compartment also had a window on each side and one at the top, and that window is visible from this angle. *Roger Freeman collection*

Armorers are preparing to load bombs onto the pylons of P-38s, including the Droopsnoot in the foreground, at a base in Belgium around March or early April 1945. On the bottom of the bombardier's compartment is a curved entry/exit door. At this late stage of the war, Droopsnoots and their formations often flew ahead of US armored columns, ravaging German forces and strongpoints. *National Archives*

"Colorado Belle" was the nickname of this natural-metal-finished P-38J Droopsnoot. It was based on the airframe of P-38J-15-LO, serial number 44-23151. *National Museum of the United States Air Force*

The left side of the pilot's nacelle or fuselage of "Colorado Belle" is seen close-up. *National Museum of the United States Air Force*

This P-38J Droopsnoot, coded F5-I, was assigned to the 428th Fighter Squadron, 474th Fighter Group, 9th Air Force. Note that the bombardier's side windows were taller at the front than at the rear. *Roger Freeman collection*

This P-38 Droopsnoot had a different design of clear bombardier's nose than is normally seen in vintage photographs, with a five-spoked metal frame, and the bombardier did not have the benefit of side windows. *National Museum of the United States Air Force*

At an open-air assembly area in the Pacific theater, mechanics are assembling P-38J Lightnings that have been shipped to this point. In the left background are two large, wooden frames for hoisting equipment. *National Archives*

A group of Lockheed Lightnings being reassembled at Nichols Field on Luzon Island, Philippines, around April 1945 includes in the foreground an F-5F photoreconnaissance Lightning, identifiable by the angular fairing on the bottom of the nose for the vertical cameras, and the raised frame for the window for the oblique camera on the side of the nose. *National Archives*

Mechanics of the 443rd Fighter Squadron have rigged a parachute over the P-38 they are working on to protect them from the intense rays of the sun. The site was an airfield at Lingayen, Luzon Island, Philippines, on April 29, 1945. The plane was P-38L-5-LO, serial number 44-25880. *National Archives*

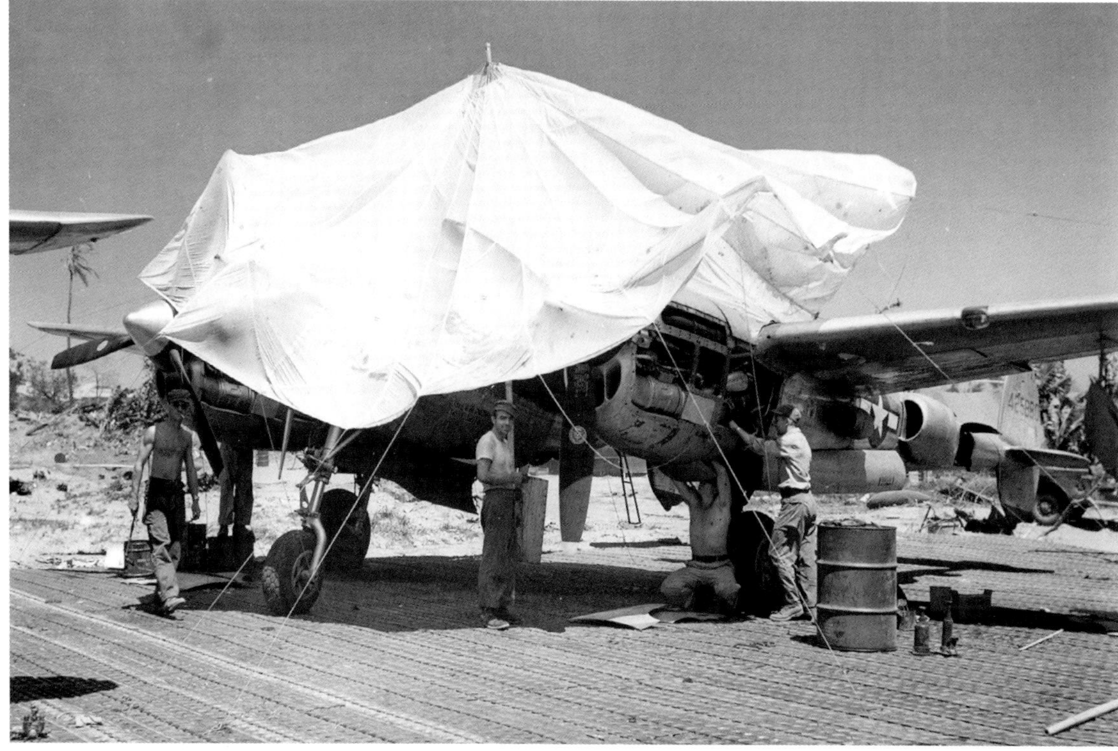

Napalm, jellied gasoline that formed the basis of a type of incendiary bomb that could spread a wide swath of destruction, was developed during World War II. It was used frequently in the war against Japan. Here, personnel of the 894th Chemical Company, 3rd Bombardment Group, are pumping napalm solution from a 55-gallon drum into a napalm tank shackled to the pylon of a P-38 at Elmore Field, Mindoro, Philippines. *National Archives*

This P-38L serving with the 475th Fighter Group, based at Lingayen Airfield on Luzon on August 25, 1945, has a lopsided mix of drop tanks. A big 310-gallon tank is under the right wing, while a 165-gallon type is on the left pylon. Because of its size and mass when loaded, the 310-gallon tank has braces, the tops of which are fastened to the wing. Another P-38L with the same complement of drop tanks is parked in the left background. *National Archives*

In Paris after the end of World War II, P-38J-25-LO, serial number 44-23652, is being reassembled below the lower arches of the Eiffel Tower for a display of US 9th Air Force aircraft that opened on August 1, 1945. The wing of the Lightning is on the truck to the right. Next to the P-38J is a Republic P-47D Thunderbolt. *National Museum of the United States Air Force*

Civilians and military men visiting the exhibition of American fighter planes at the Eiffel Tower in August or September 1945 include a number of them who are on a platform and stairs over the right-center section of the wing of P-38J-25-LO, serial number 44-23652. The platform allowed the attendees to view the cockpit of the Lightning. Behind the P-38J and the neighboring P-47D was a concourse containing an exhibition of photographs of the recent air war. *National Museum of the United States Air Force*

Around the end of World War II, American servicemen survey the carcasses of damaged and destroyed P-38 Lightnings at a 7th Air Force base in the Pacific theater. Whenever possible, parts were salvaged from such aircraft to enable other planes to continue to fight. Parked in the background are a number of B-24 Liberator bombers. *National Archives*

After the cessation of hostilities in World War II, the US armed forces, particularly in the more remote areas of the Pacific, frequently scrapped their warplanes rather than expend the considerable effort and expense of shipping the aircraft back to the United States. Planes often were unceremoniously deposited in scrap piles such as the one shown here, strewn with Lockheed P-38s. This was an ignominious end for a class of warplanes of unmatched grace and combat effectiveness. *Stan Piet collection*

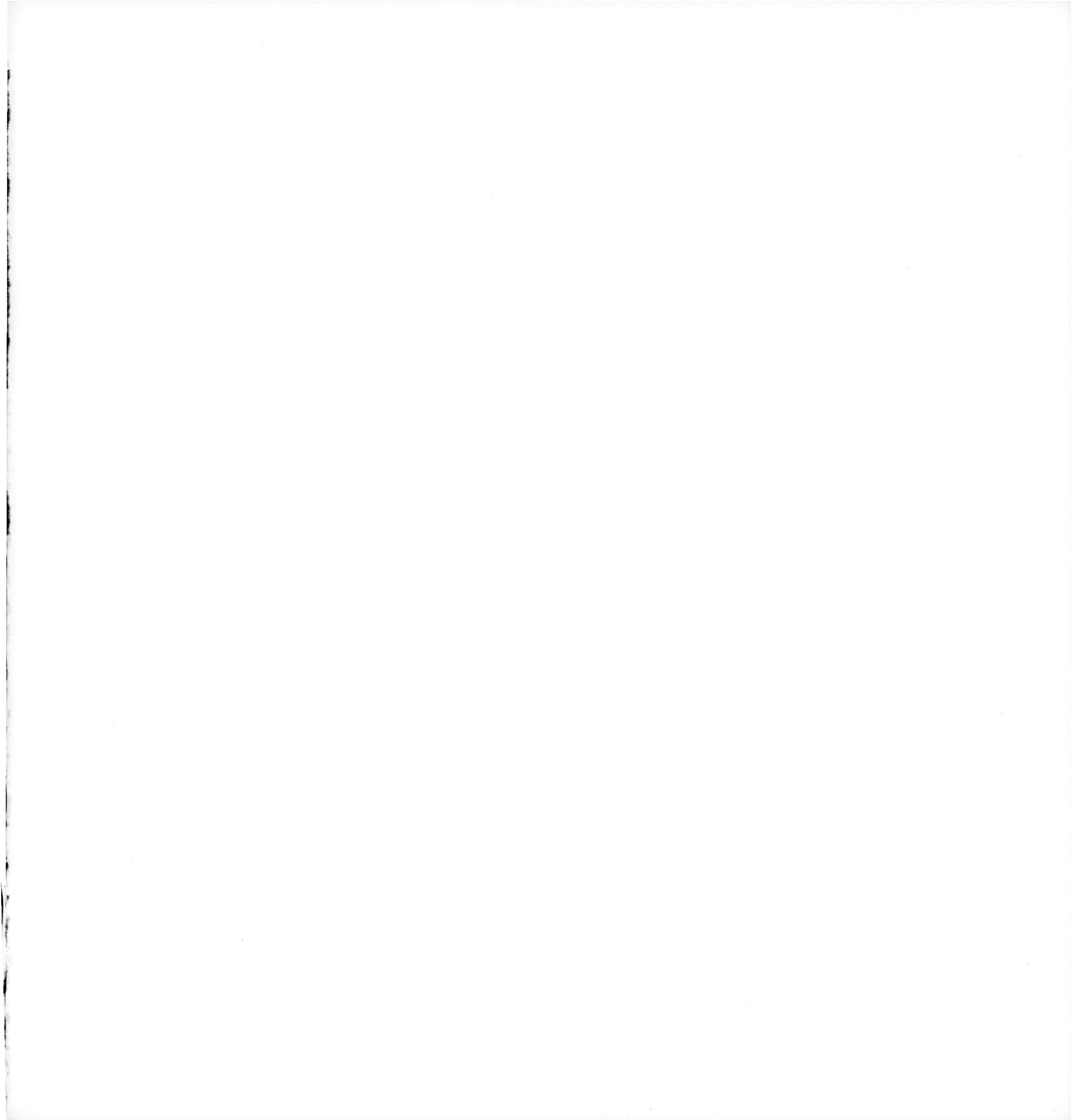